QUANTUM

PHYSICS

FOR BEGINNERS.

AN EASY GUIDE TO UNDERSTAND THE BASIC CONCEPT AND PRACTICAL APPLICATIONS OF QUANTUM THEORIES.

JOHNSON BRIGHT.

TABLE OF CONTENT.

INTRODUCTION.

The study of matter and energy at the bulk vital level is known as quantum physics. It goal is to reveal the attribute and behaviors of the very building blocks of nature. Although many quantum experiments examine very small objects, such as electrons and photons, quantum phenomena are all around us, acting on every scale.

This is also physics of the microscopic world, the world of molecules, atoms, and sub-atomic particles. It is a branch of physics that was born in the early decades of the twentieth century, with its seeds germinating in the late decades of the nineteenth century.

Quantum physics was born out of a mixture of failure to understand certain experiments, and also out of non-sense results when applying classical physics to atoms. Yes, physicists were already familiar with the notion that matter is composed of microscopic packets, the atoms. This notion was initially put forth by the greek philosopher democritus who postulated that everything is built from atoms and the void. Later on, chemists starting in the eighteenth century with english chemist john dalton, and after carefully studying chemical reactions and the proportions of reagents that go in and the proportions of the resultant chemicals, correctly inferred that matter is built of individual packets, which he called atoms, dmitri mandeleev in the nineteenth century classified atoms by their chemical properties and arranged them into the periodic table of elements that we are familiar with today

.CHAPTER ONE.

QUANTUM PHYSICS VS QUANTUM MECHANICS.

- What is quantum physics?
- Why should we care about knowing what is quantum physics?
- Do we need to know what quantum physics is?
- How quantum physics does affects our daily lives?

QUANTUM PHYSICS.

This is a physics of the microscopic world; the world of molecules, atoms, and sub-atomic particles. It is a branch of physics that was born in the early decades of the twentieth century, with its seeds germinating in the late decades of the nineteenth century. Quantum physics was born out of a mixture of failure to understand certain experiments, and also out of non-sense results when applying classical physics to atoms. Yes, physicists were already familiar with the notion that matter is composed of microscopic packets, the atoms. This notion was initially put forth by the greek philosopher democritus who postulated that everything is built from atoms and the void. Later on, chemists starting in the eighteenth century with english chemist john dalton, and after carefully studying chemical reactions and the proportions of reagents that go in and the proportions of the resultant chemicals, correctly inferred that matter is built of individual packets, which he called atoms, dmitri mandeleev in the nineteenth century classified atoms by their chemical properties and arranged them into the periodic table of elements that we are familiar with today. You will see as go on with our story that the chemical properties of atoms are a direct consequence of quantum physics, physicists knew that matter is overall neutral, which is made up of equal of positive and negative charges. So after verifying that matter is made up of atoms, experiments focused on understanding the architecture of the atom. Basically, understanding how the negative (electrons) and positive charges (protons) are arranged in atoms. Enter new zealander physicist ernest

Rutherford in the first decade of the twentieth century who shot alpha particles, helium atoms that that have lost their two electrons, onto thin gold foil. Rutherford observed that the alpha particles (positively charged) were deflected from positively charged point particles, of much higher charge. Atoms are made up of dense positively charged nuclei surrounded by orbiting negatively charged electrons. All this is fine and nice and a source of confidence, so much so that many physicists at the beginning of the twentieth century were predicting the death of physics. Their opinion went something along these lines: "It is only a matter of a few years when we know all that we need to know about the fundamental physics of nature. We know everything. It is now just a matter of applying our physics to the world around us. And that is exactly what they did, but nasty surprises awaited them! but nasty surprises awaiting scientists even today, just a decade before Rutherford conducted his experiments, german physicist max planck was studying radiation emitted by the so-called blackbody.

- Blackbody: In physics, a blackbody is an object whose peak emission wavelength, or in other words color, is related to its temperature. Imagine a piece of iron that you throw in a fire, as it gets hotter it first glows red (long wavelength), and when it gets even hotter it glows even brighter but its color will become blue (short wavelength). This piece of glowing piece of iron is referred to by physicists as blackbody. What happens if one were to increase the temperature of a blackbody even further, that is what planck was thinking about., the blackbody should glow yet brighter but in the ultra-violet was predicted by the relevant physics known at the time, thermodynamics. The color of the blackbody should not only shift toward shorter and shorter wavelengths, but its brightness should become increasing higher very fast, so fast that it reaches infinity! But his is non-sense. In what manner can warming an object, primarily pulling out a finite amount of energy into it, make it emit an infinite amount of energy? This is prohibited by the law of conservation of energy. Planck's success was postulating that radiation, or light, is quantized in individual packets of energy, called photons. The energy carried by each photon is inversely proportional to the wavelength of the radiation, with the constant of proportionality now called planck's constant. None other than the albert einstein used this result to describe the photoelectric effect experiment (another important experiment of quantum physics), for which he later won the nobel prize. Other physicists in the early decades of the 20th century applied their physics known to them at the time classical mechanics and electromagnetism to the simplest of atoms, the hydrogen atom (one electron orbiting a single proton).

- Classical mechanics: it's says that an object orbiting a center must experience centrifugal acceleration.
- Electromagnetism: this says that an accelerating charge must give off radiation (energy).

Now, let us put one and one together. The electron, a charged particle, as it orbits the proton in the hydrogen atom, will experience acceleration so it must radiate. So far so good. However, the radiation emitted by the electron will steal away its energy making its orbit decrease and spiral toward the proton in a fraction of a fraction of a fraction of a second! So basically, classical physics predicts that no atoms should exist and by consequence, we too should not exit! This non-sense result was a slap in the face of early 20th century physicists that all not is well with physics, or at least there must be something over and beyond the classical physics known till then:

They are classical mechanics, thermodynamics, electromagnetism, and optics. This new physics became known as quantum physics. Its cornerstone is that all physics quantities that could be observed, the so-called observables, are all quantized. That is they come in individual, discrete units instead of taking on a continuum of values, as in classical physics. Faced with the above "atoms should not exist debacle," Danish physicist neil bohr proposed in the second decade of the 20th century his toy model of the hydrogen atom. He postulated that the existence of stationary orbits in which the electron, if it satisfies a certain condition, does not radiate its energy away. That condition is related to its orbital angular momentum and it dictates that the orbital angular moment of the electron in this orbit must be a definite multiple of Planck's constant.

This toy model was able to explain to a high degree the spectrum of emission of hydrogen atoms observed in the lab. Such work of theory and experiment going hand-in-hand was the way in which quantum physics was built. Moreover, quantum physics was a group effort in which nearly all physicists working at the time contributed to the endeavor, some more than others, of course. This kind work continued well into the 20th century in which quantum theory was fully development and compared against experiment. Of course, needless to say, experiment (which is science's truth-o-meter) agreed with the predictions of quantum physics.

The utmost achievement theory in physics is quantum theory. It has also spawned new fields: materials science (think smaller and more power computer chips, led tvs, solar photovoltaic cells, etc), nuclear physics (think nuclear power and nuclear medicine), and particle physics (theory to describe the sub-atomic world and related to how the Universe came to exists. Quantum theory at the utilitarian level is very powerful and leads to practical results that permeate our everyday lives. However, problems with quantum theory sprung up from the very beginning. These problems are at the interpretation level: how do you interpret the story told by the equations. Keep in mind that the equations themselves are correct and self-consistent; they give specific, well-defined predictions that turn out to be correct when tested experimentally. The problems arise when you try to understand the equations! neil bohr remains to the day one of the giants of quantum physics and the father of the most intriguing interpretations of quantum physics. An interpretation so bizarre but also undeniably true that it gave albert einstein many a headache. This interpretation is known as the copenhagen interpretation.

In a nutshell, the Copenhagen interpretation stipulates that if I don't see you, then you don't exist. At the fundamental level, it states that reality depends on the observer! The role of the observer is paramount in quantum physics. The observer and the act of measurement, such as "seeing," itself affects the outcome of the experiment. The reason for this is relatively easy to understand. Imagine going into a dark room holding an unlit flashlight. Turning the flashlight on will shoot photons from it to an object in the room, a chair for example. These photons bounce on the chair back to your eye and that is how you see the chair. Now, the energy carried by the photon is very small compared to the energy that could move the chair and the chair remains in place. So the act of measurement here did not change our system, the chair. Now imagine that instead of a flashlight, you have a gun that shoots golf balls at very high speeds. You go into the room and start spraying with fast golf balls. One or two golf balls are probably going to hit the chair and maybe also probably move it too. From the sound of the golf balls crashing on the chair you will let you know where the chair was located, but now the new position of the chair becomes unknown. Therefore, the act of calculating here changed the system. Of course, in quantum physics, all systems are microscopic and the measurement tools are also microscopic leading measurements to messing with the system.

The Copenhagen interpretation takes this notion to the max and says that the very act of measurement makes a system exist! One might be tempted to say, this is just an interpretation and it need not be correct. However, recent experiments done at top quantum optics labs strongly suggest that the copenhagen interpretation is correct. A vary graphic illustration of the apparent absurdity of quantum physics and the copenhagen interpretation is the schrodinger cat paradox. In this famous experiment, named after irish-austrian physicist ewrin schrodiner, a cat is put in a closed box with a sealed vial of poison. A radioactive source might decay at random and emit a radioactive particle that would break the seal thus killing the cat. To an outside observer, the cat has 50-50 chance of being dead or alive. The way this is expressed in quantum physics formulas, the so-called state function, will assign equal weights to the cat being dead or alive. A physicist can do all calculations required and compare to experiments and all will be fine, but when the physicist tries to interpret the state function formula, all hell breaks loose: the equation seems to be saying that the cat is dead and alive at the same time! But remember, who determines the truth? It is the act of observation itself: the measurement. The observer will have to open the box to determine whether the cat is alive or dead. Another weird aspect of quantum physics is the so-called particle-wave duality, or the complementarity principle. In our macroscopic world, matter behaves like particles (the ball is here, the bike is there) and light behaves like a wave (cause of the rainbow seen on a dvd's data surface). In the microscopic world governed by quantum physics, an entity (matter or light) will behave as a particle or a wave, depending on the experiment (more support to the Copenhagen interpretation above). French physicist Louis de Broglie, author of probably the shortest doctoral thesis in physics in which he presented the smallest formula that would later win him a nobel prize in physics, proposed in that same thesis in the nineteen twenties that a particle with a given value of linear momentum, basically mass multiplied by velocity, will have the wavelength of its complementary wave inversely proportional to its momentum. The constant of proportionality is planck's constant, yet again! Experiments done on electrons showed that they behave like a wave upon passing through very narrow slits, a behavior seen for light also.

Albert einstein successfully explained the photoelectric effect, mentioned before, by invoking the particle nature of light: in that experiment light will behave like small particles, photons.

QUANTUM MECHANICS.

You are currently facing one of the most important equations of all time. It is called the Schrödinger wave equation. Let me explain why it is so beautiful. First of all, as you can see, it is really simple. Usually, in the physics world, simple and elegant formulas are the most important ones. Even einstein had this strong belief that the world and the universe could have been described by means of some...cute formulas. In terms of equations a well-built theory is usually visually good. As einstein himself said that, it can hardly be contravene that the highest aim of all theory is to make difficult basic elements as simple and as few as feasible without having to surrender the sufficient illustration of a single datum of exposure. "Just think about it. It's like when you want to express your emotions. Sometimes you don't think too much about them, your words come out of your mouth and you produce a messy cascade of words. Your interlocutor can still understand what you are saying, but it is kind a difficult to follow your thoughts. But if you write your thoughts on paper, you will probably find a better way to express them. A more efficient and concise way. This also holds for physics theories. We could have good ones, but some of them are just more elegant than others. Of course, this is a hard job, and our scientific inability to simplify is something we should always accept. But besides its formal beauty, the schrödinger equation tells us something more. It is the starting point for the understanding of quantum mechanics. What is it? i will explain it to you, keeping it simple. I could start and end this tutorial by saying that quantum mechanics is the study of very small things, but you would be very disappointed then. So i will try to reveal something more to you. First of all, for "very small things" we mean we are interested in stuff that exists in the real world, but has atomic-scale dimensions. We are talking about atoms and subatomic particles. So we can say quantum mechanics deals with the atomic and subatomic world. And if you take a lot of particles, you have the macroscopic world.

The macroscopic one, is the world you, me, we are used to. In everyday life, we have to deal with macroscopic objects. Your moka is a macroscopic object. However, it is made of atomic and subatomic particles. Now the macroscopic world is well described by Classical Physics, for example, Newton gave us some laws that fit well what we observe in the everyday life.

CLASSICAL PHYSICS.

This can help us understand why the earth orbits around the sun, why we have seasons, how planes fly and much more. So it is really useful. But at a certain point, at the windup of the 19th and the outset of the 20th century, scientists realized that something was missing. When they decided to study the atomic and subatomic world, they found classical physics didn't work anymore. They needed another approach. This was a huge issue, a problem that needed to be solved. In fact, Physics is no more physics if it can't describe reality. The passion to settle inconsistencies between observed phenomena and classical theory led to two major revolutions in physics that created a shift in the original scientific paradigm, the theory of relativity and the development of quantum mechanics. The most important result is that light behaves in some aspects like particles, and in others like waves. I know what you're thinking: no way! And that's the same thing physicians were thinking when they first approached the undiscovered world of quantum mechanics. Let me explain it better.

- Matter.

The "stuff" of the universe, consists of particles such as electrons and atoms. But it also exhibits wave like behaviour. This exposure has been confirm not only for a simple particles but also for compound particles like atoms and even molecules. For particles that are macroscopic, since their extremely wavelengths is short and their plus six wave properties usually cannot be detected. Although in physics, the use of wave-particle duality has worked well, the concept or explanation has not been satisfactorily resolved. Bohr called it the "duality paradox" and regarded it as a most important or ontology fact of nature. A given kind of quantum object will show from time to time wave, a particle and character, in sequentially different phenomenal settings. He viewed such duality as one features of the idea of complementarity. Talking about this wave-particle duality, Einstein said: "It seems as though you must use from time to time the one theory and the other, while at the same times you may use either. You are faced with a fresh kind of obstacles. You will have two inconsistent pictures of reality independently neither of them fully analyze the significant fact of light, but they do it as one. The most unique experiments that is famous and allowed scientists to understand the dual nature of matter was the double-slit experiment.

DOUBLE-SLIT EXPERIMENT:

This experiment shows, with unparalleled abnormality that slight particles of matter have something of a wave about them, and suggests that the very act of observing a particle has a dramatic effect on its behaviour. To start off, imagine a wall with two slits in it. Imagine toss out tennis balls on the wall, a few will bounce off the wall, but a few will travel through the slits. If there's another wall behind the first, the tennis balls that have travelled through the slits will hit it. If you mark all the spots where a ball has hit the second wall, what do you expect to see? That's right. Two strips of marks roughly the same shape as the slits. In the image below, the first wall is shown from the top, and the second wall is shown from the front. (Show some double-slit experiments) Now imagine a light at a wall with two slits. As the wave passes through both slits, it essentially splits into two new waves, each spreading out from the slits. These two waves interact with each other, and they are said to interfere with each other. The interference could be disruptive or constructive, and in the first case, they will cancel each other out. In the second case, they will reinforce each other, giving spots with the brightest lights. So as soon as the light come across the second wall situated at the back of the first, you will see a stripy pattern called an interpose pattern. Now, if you do the same thing with a beam of electrons, you would expect to see two rectangular strips on the second wall, as with the tennis balls, because they are particles. But what you actually see is that the spots where electrons hit replicate the interference pattern from a wave. As you can see, this experiment suggests that what we call "particles", such as electrons, somehow combine characteristics of particles and characteristics of waves. And this is the real essence of the quantum world.

Basically, everything can be described or associated with a so-called wave function. In physics, we indicate a wave function by means of the Greek letter "psi": ψ and if you remember, the psi function appears in schrödinger's equation.

In quantum mechanics, the Schrödinger equation is a fundamental equation that determines the temporal evolution of the state of a system, for example of a particle, an atom or a molecule. When it comes to quantum mechanics, intuition is no more present. For example, if you hold a ball you will notice it has some mass because of its weight. You can feel its weight, and if you take something heavier you will notice. It is just something very intuitive, and the classical physics world is built upon this kind of intuition. However, when it comes to quantum physics, things get more complex, and we soon realize that we can't predict exactly what is going to happen to the motion of a ball, for example. Or we can't really tell, for example, if a cat is black or white. One has to imagine reality as a set of possible configurations, and you don't know a priori which configuration is going to be chosen. We can tell that there is a certain probability associated with each configuration. Only when we make measurements, we can see the chosen configuration, which we call state. Let's go back to the cat. Let's suppose you want to pet him. We have a cat, but we don't know if it is black or white or whatever colour. Quantum mechanics essentially states that the only way for us to know which colour the cat is, is to "measure" it, and we do so by applying some mathematical objects to the "state" of the cat, which is called "operators" you have a wave function, a state (which in this case is the colour of the cat), you apply an operator, you get the result black, white or anything at all. If you repeat the experiment a high number of times, you will end up with a distribution probability of colours, and this distribution will have a peak on one colour, and that will indeed be the colour of the cat you are want to pet.

One of the requirements of quantum mechanics is the description of macroscopic reality. In the "big number of atoms" limit, quantum mechanics should reproduce the macroscopic world. That is, putting in the crassest possible terms, if you see the cat is white, quantum operators applied to the colour configuration should mainly give the "white" result. Numerous features of quantum mechanics are illogical and can seem to be convoluted due to they describe behaviour quite different from that seen at larger scales.

What did quantum physicists richard feynman said about quantum physics? Feynman said that quantum physics deals with "nature as she is absurd".

For example,

- The uncertainty principle of quantum mechanics

means that the more closely one pins down one measurement (such as the position of a particle), the less perfection of another collective measurement applicatable to the same particle (such as its speed) must become, If you know very well the position of a particle, you will lose a lot of information about its speed. And vice-versa, you know very well the velocity of a particle, you will keep losing track of it.

- Entanglement, in which a measurement of any

two valued state of a particle (such as light polarized up or down) made on either of two "entangled" particles that are very far apart causes a subsequent measurement on the other particle to always be the other of the two values (such as polarized in the opposite direction).

IMPORTANCE OF QUANTUM MECHANICS.

> ➤ It teaches us an important thing about nature.

Its description is essentially probabilistic. Before quantum mechanics, we were used to thinking that the world is governed by some given laws that give precise results. You act, you get precise results, and every action is followed by a reaction that seems to be predictable. Instead, nature doesn't work like that. The probability of an event for example, where on the screen a particle shows up in the double slit experiment is related to the square of the absolute value of the amplitude of its wave function. We can just say, for example, that we have an 80% of chance that we'll find the particle in a given position interval, but we will know where it is only when we will measure it.

➢ It tells us is that is not possible to know the values of all of the properties of a system at the same time. This is of course due to the uncertainty principle we discussed before.

➢ Last but not least, in quantum mechanics is a good theory because, even if it studies the "very small things", it closely approximates the classical description of nature in the case of large systems.

CHAPTER TWO.

A BRIEF HISTORY OF QUANTUM MECHANICS.

Traveling to the 1750s, we find out that the scientists we're putting different substances in flames and passing the resultant light through a prism. They found that the hot gas is given off by the burning materials emit different colors of light or spectrum. For example, ordinary table salt generated a bright yellow spectrum, furthermore not all the colors of the rainbow appeared there were dark gaps in the spectrum. In fact for some materials there were just a few patches of light.

➢ The element.

More than 1820s spectra was recognized in the sense that it provided an excellent way to detect and identify small quantities of an element in a powder that was put into a flame, at the same time the white light of the sun was also being examined closely. It was find out that the solar spectrum itself had tiny holes in 1802, there were a few thin dark lines in the rainbow of colors, but the reason for spectral lines in the light and the relationship to each substance was a real mystery. Traveling forward from that era to a little over a hundred years ago, scientists were examining the colors of light given off by solid heated objects, they discovered that these hot solids gave off continuous spectra and that the overall color of the light revealed the temperature of the object.

Now this was important because scientists realized that this discovery made it possible to measure the temperature of an object from a distance, they could measure the condition of the sun for example. During these discoveries they also notice that some of the objects absorb light extremely well almost perfectly, in fact, they would called black bodies because they absorb virtually all the light that was shone upon their surfaces, these same objects also radiated almost perfectly and as noted before the temperature of the blackbody object determined the distribution of colors or wavelengths in the admitted light.

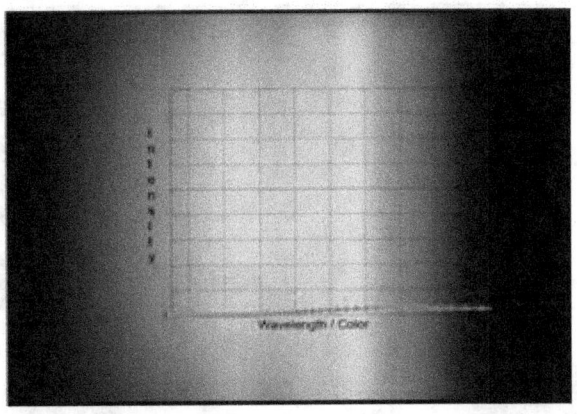

This curve (as shown in the figure above) shows how much light of each color is admitted by a cool object. And there isn't much light and what light there is mostly lies out past the red end of the spectrum in the infrared. the figure on the right will show that different colors added together as we progress, right now only red light is visible, as shown in the figure below,

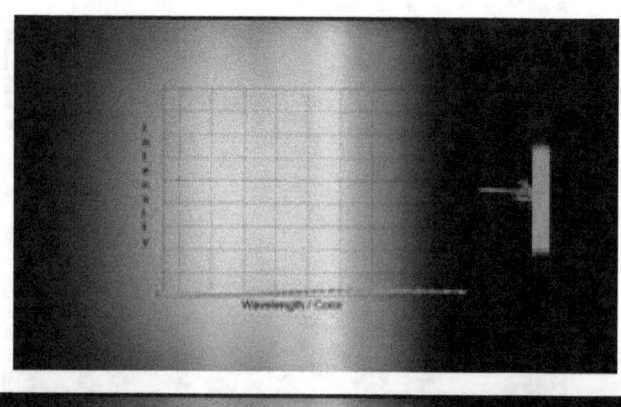

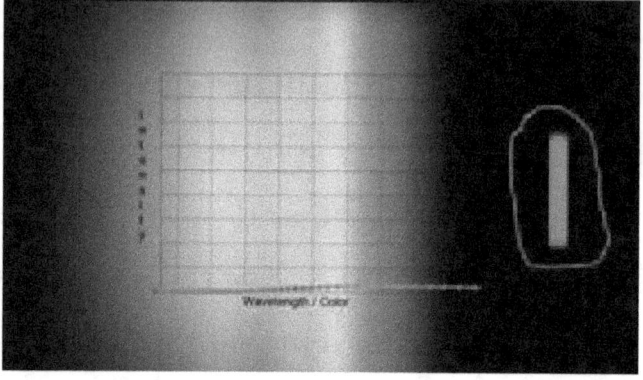

The below curve shows how much light of each color is emitted by a medium temperature object as shown in the figure below,

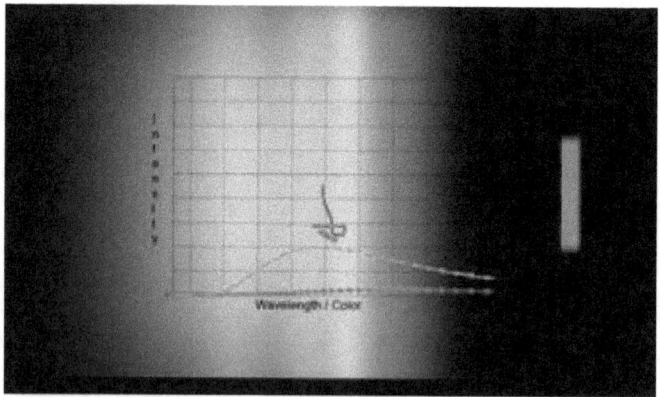

and the most light is emitted in the orange, yellow, green wavelengths, so now that we add orange, yellow and green to our cauldron of light on the right and as you can see the combinations so far looks yellow as shown in the figure below,

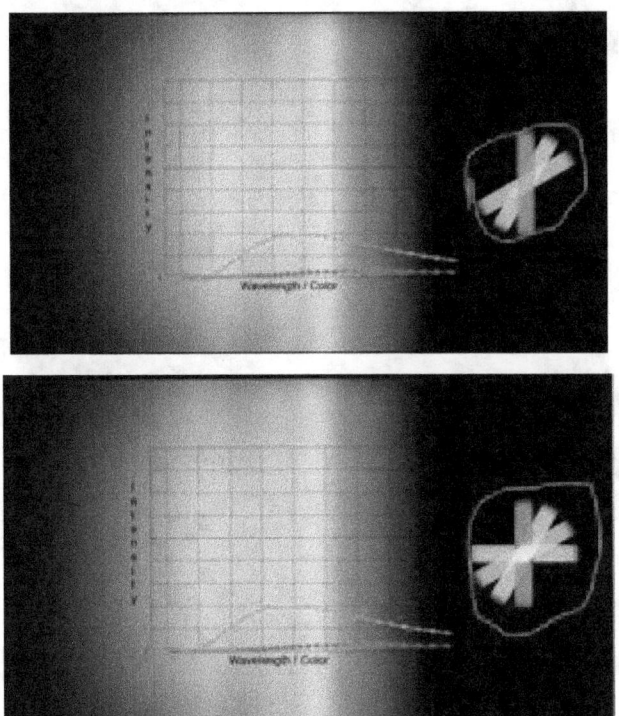

the third curve is for a really hot object as shown in the figure below, lots of light and with most of it being emitted toward the blue end of the spectrum as shown in the figure above,

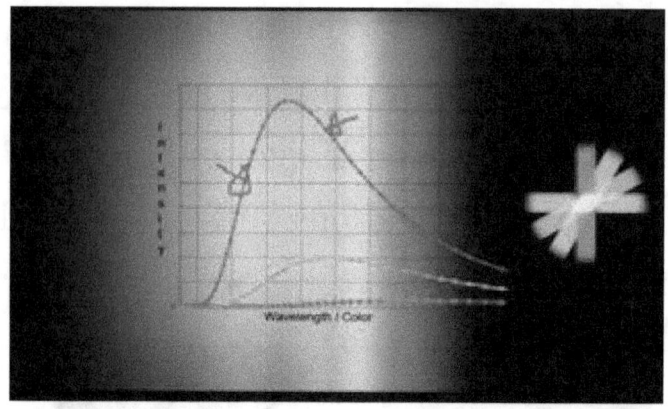

and as you can see our cauldron now is pure white in the center where all the colors overlap as shown in the figure below.

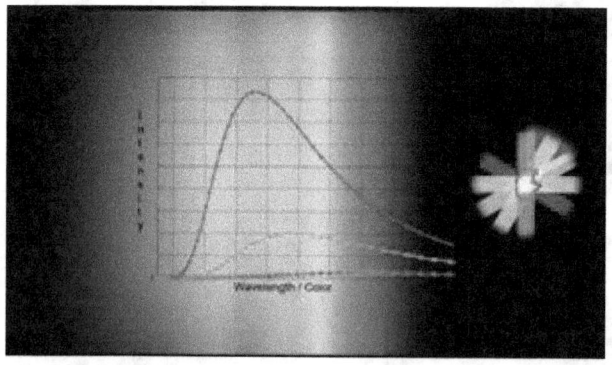

A heated blackbody follows the color path as the temperature rises. A good example of blackbody radiation is the heat inside a kiln.

Inside the kiln, electromagnetic radiation light and heat exist in a form of standing waves. Waves that like the vibration of a guitar string are attached at both ends, the ends of the standing waves like the ends of the guitar string do not move, they are anchored to the sides of the kiln. And many waves exist with varying wavelengths of color.

At low temperatures the primary color inside the kiln is the infrared and we cannot detect it with our naked eye, but as the temperature rises the kiln begins to glow red and as the temperature continues to rise, the dominant color changes to orange then yellow then bluish white.

The distribution of energy in light shifts to shorter wavelengths as the temperature rises, but there was a problem the scientists expected the distribution of emitted light to continue to the increase at wavelengths toward the ultraviolet end of the distribution, it didn't instead there was less and less light given off as they were further and further into the ultra violet, this was called the ultraviolet catastrophe.

But it wasn't a catastrophe at all, it was the beginning of something remarkable a truly great scientist named max planck soon figured out a way to explain the observation he concluded that the energy contained in the standing waves inside the kiln did not and could not possess just any and all different amounts of energy, instead the quantity of energy the standing waves possess had to be limited to a few specific discrete values of energy for each color. A standing wave of blue light for example can have energy equal to zero electron volts or three electron volts or six electron volts or nine electron volts and so on. In general energy equals to any whole number n times plunks constant h times the frequency of light nu as shown in the figure below,

which for blue light is any number times three electron volts, but notice blue light standing waves cannot contain energy equal to one electron volt or two electron volts or four electron volts, realizing that energy could have only discrete values was the beginning of quantum mechanics. The energy is said to be quantized and n from the equation above is called a quantum number.

plunks conclusion that light energy is quantized was quickly used by Einstein to explain another puzzling phenomenon, it was known that to shine a light upon a metal plate can release electrons from the plate, but the light has to have a certain wavelength before a single electron is released, we can shine brighter and brighter light on the plate forever but if the color isn't right the electrons stay home.

Einstein concluded that the light striking the plate had to be coming in discrete bundles and unless a single bundle had enough energy to free an electron from its captivity, it would remain trapped and as plunk has suggested the wavelength and frequency of the light was a measure of the amount of energy each bundle carried, so while blue light packets might be able to free an electron, red light packets could not no matter how many red packets hit the plate. but that sounds awfully like light is a particle and not a wave and there are mountains of data showing light behaving like a wave, diffraction, refraction, interference, which is it a particle or wave, it is both. These light packets are extremely tiny, let's call them photons from photons the Greek word for light and when you try to explain the behavior of things that is small, you must resort to really unusual ideas, the idea is contained in quantum mechanics.

QUANTUM ENTANGLEMENT.

This is one of the utmost intriguing and perplexing phenomena in quantum physics. This permit physicists to form bond between particles that seem to violate our own knowledge of space and time.

But what is entanglement, really?

These two coins are not entangled as shown in the figure below,

If you flip this coin as shown in the figure above, it won't have any effect on the other. Each coin, when flipped, it will have a 50% chance of landing heads or tails, and

that's true no matter what you see when you look at the other. But what if we could do some magic trick to entangle these coins? Sort of-one comes up, heads, the other will come up tails. No matter how far apart they are. For example, let's say you give this coin as shown in the figure above to Alice and the other one to Bob just before they set off on rocket ships going in opposite directions. Alice and bob are each under strict instructions not to flip their coins until they've been traveling for ten years. When the time is up, they each flip their coin and immediately send radio signals to each other with the answers. But while those signals might take years to reach their destinations, the entanglement means that the moment alice sees heads, she knows that bob's coin landed tails and vice versa. Alice doesn't have to wait for bob's radio signal to arrive to know what it will say. Alice's coin had a 50/50 chance of coming up heads. But once it does, bob's coin is tails 100% of the time. Because bob's signal is still traveling to her when she finds out the answer, it's as if Alice has predicted the future.

This might sound impossible, but it's exactly what quantum entanglement does to individual particles in experiments. Let's say you carefully prepare two particles in a lab to make them entangled. One way to set this up is to ensure that they will have opposite spins, meaning that they are lineup opposite to each other in a magnetic field. Then, you separate the particles by thousands of miles and measure the spin of one of them. If you find that its spin points up, the others will always point down. Like in the coin example, whenever alice measures the spin of her particle. If it has spin up, she'll instantly know that bob's particle has spin down or vice versa. No matter how far away they are.

Einstein called this spooky action at a distance. Because according to his theory of relativity, no information can travel faster than the speed of light. You shouldn't be able to know anything about the other particle before a light beam with the information can reach you.

Entanglement seems to throw this idea out the spaceship airlock. To be clear, however often it's described that way in science fiction, this is not faster than light communication. Just like with the coin flip, the state of the entangled particle in your possession is random. You can't know ahead of time if it's spin up or spin down. All you know is that when you measure it, the other particle will be the opposite. Even if Alice and Bob tried to agree ahead of time on some kind of code to use for a message, they wouldn't get to decide what message they were going to send. It would just be random gibberish.

WHY QUANTUM ENTANGLEMENT CAN'T BE USE TO COMMUNICATE INSTANTANEOUSLY.

It can use it in technologies such as quantum cryptography and quantum computing, but how does it work? How do the particles know what to do? Are they passing messages that we can't see? And why can you entangle particles but not coins? And that is one of the most interesting and puzzling concepts of quantum mechanics.

In fact, Einstein was one of the first one kind of dabble in trying to understand it, in fact he wanted to show that it was not part of our world. And then the quantum mechanics is not the correct theory, but his discussion was very wordy and kind of conceptual and not very practical. And it took another 20, 25 years before an

Irish physicist, john bell, kind of turned his ideas into an equation. If entanglement didn't exist, the equation would be satisfied. If entanglement exists, it would be violated. The world is quantum mechanical. This is absolutely amazing. This is something we cannot understand with our usual concept of classical mechanics. But it is there. It exists.

Quantum entanglement is certainly strange and we still don't completely understand it. But it is a feature of nature and it's an incredible tool that we can use. The results of these experiments have applications in new technologies that will forever change our world.

CHAPTER THREE.

QUANTUM FIELD THEORY.

Physics is a wonderful way for people to understand the world around them. It can explain how birds fly, why ice freezes, what makes fire glow, just to mention a few things. Now, this doesn't meant that the explanations are always easy and clear. In fact, the explanations can be downright murky for phenomena far from the familiar. Perhaps the two physics theories that are the hardest to accept when you first encounter them are Einstein's theory of relativity and quantum mechanics.

The Einstein's theory of relativity talks about how moving clocks run more slowly than stationary ones and how objects in motion appear to shrink.

While the second one which is quantum mechanics tells us that no measurement is certain and that probability reigns supreme in the subatomic world. While I'd like to tell you that there is some credible debate about these theories and that that we scientists have somehow figured out a way to return to the more intuitive physics of the late 1800s, but that isn't true.

The simple fact is that relativity and quantum mechanics have been tested countless times and they work. Like it or not, you have to learn to accept these weird predictions are just, simply, well, true. However- and this might blow your mind- these ideas are also about a hundred years old. Frontier science has actually moved on.

What scientists currently think is even weirder still. So let's review a bit what traditional quantum mechanics is all about. It was invented in the mid-1920s and it was exemplified by an equation devised by Erwin Schrodinger, what we now call Schrodinger's equation. And it is shown in the figure below.

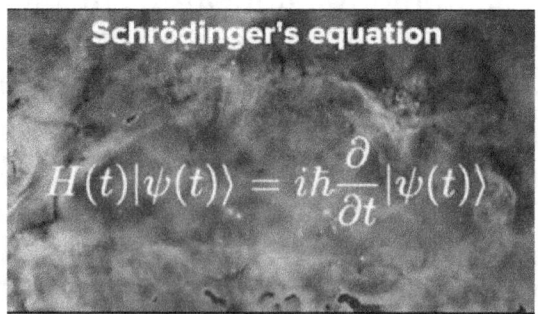

$$H(t)|\psi(t)\rangle = i\hbar \frac{\partial}{\partial t}|\psi(t)\rangle$$

This equation explained why electrons had only certain energies and positions when they circled an atom. At its very simplest, the equation explained that an electron could be here or there, as shown in the figure below.

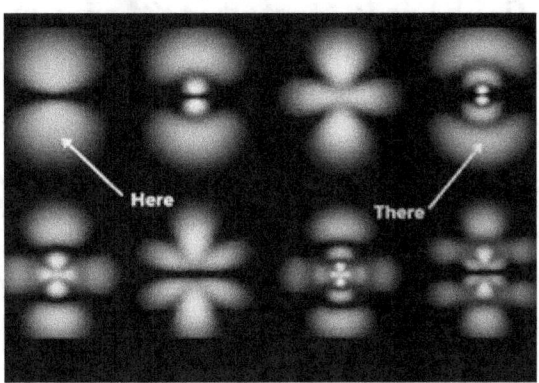

But never here nor there as shown in the figure below.

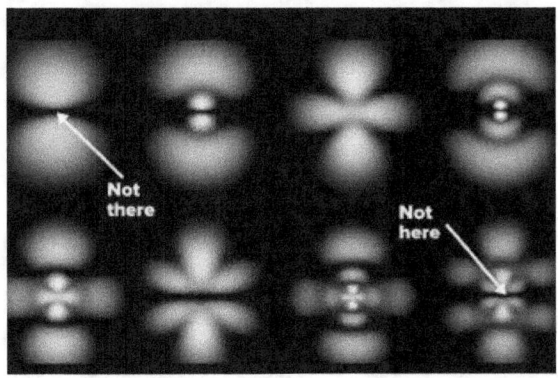

That's what "quantum" means. There are certain, discrete things that are possible and others that aren't. The things that were quantized could be mass, charge, position or energy. Now, Schrodinger's equation was only a partial quantum theory and it didn't take into account relativity. What it did was take a proton and assume it was surrounded by a classical electrical field. Remember that classical fields are not quantized. They vary smoothly. However, things changed in the late 1920s when Paul Dirac started puttering around with quantum mechanics. One thing he did was successfully merge quantum mechanics and einstein's theory of special relativity. Another thing he spearheaded was to figure out a way to make a fully quantum theory. He did this by finding a quantum formulation of the electric field surrounding the proton. And this is called the "second quantization revolution.

SECOND QUANTIZATION REVOLUTION.

This just means that the electric field was expressed quantum mechanically and that it joined such things as a quantum description of the location of matter, which was covered by the first quantum revolution. In the ensuing years, these ideas have been generalized to cover all of the subatomic forces, specifically the strong nuclear force, the weak nuclear force and electromagnetism. While each force has a different precise formulation, they are all examples of what we now call a quantum field theory or QFT. Although each theory has its own interesting peculiarities, let's talk a little about some general truths of all quantum fields.

MODERN PHYSICS THEORY.

In modern physics theory, one can picture all subatomic particles as beginning with a field. Then the particles we see are just localized vibrations in the field.

However, in accordance with quantum field theory, the proper way to think of the microscopic globe is that everywhere and I mean everywhere there are myriad of fields, up quark fields, down quark fields, electron fields, etc. And their particles are just localized fluctuation of the fields that are moving around. The idea can also explain how particles interact.

For example, suppose you have an electron moving along. The electron is a localized vibration of the electron field. If the electron emits a photon, then the quantum field theory way of looking at things says that some of the energy of the electron field sets up a localized vibration of a photon field which then moves away. So those are the essential features of quantum field theory.

Theoretical physics simply imagines that ordinary space is full of fields for all known subatomic particles and that localized fluctuation might be seen throughout. One another with can interact these fields, like two close tuning forks. These interplay elaborate how particles are created and destroyed. Primarily, the energy of some vibrations change from one field and set up vibrations in another kind of field. Now, actually calculating something with this prescription is really, really very difficult. The math can be pretty crazy. But the essential concept is really very simple. If you look around you and you have even the smallest ability to think creatively, you can imagine these vibrations everywhere you look. I don't know about you, but i think that's an awfully cool idea.

SCHRODINGER EQUATION.

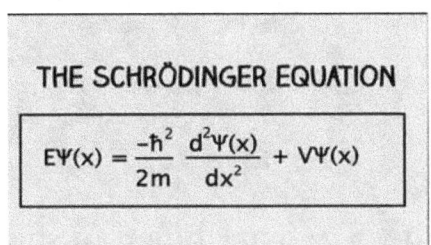

THE SCHRÖDINGER EQUATION

$$E\Psi(x) = \frac{-\hbar^2}{2m} \frac{d^2\Psi(x)}{dx^2} + V\Psi(x)$$

The Schrodinger equation is one of those things that pops up a lot in like quantum science articles and journals and stuff but the journalist doesn't usually go into what it means, which is fair because it's a fairly complex topic. So now i just wanted to share with you guys like what it actually means so next time you read it in an article you can get a better gist of what it's about.

Schrodinger equation tells us everything we can possibly know about a quantum system. It's basically the F=ma of the quantum world. If a ball is tossed up by you and solve F=ma, you can predict its position and momentum for any moment in time.

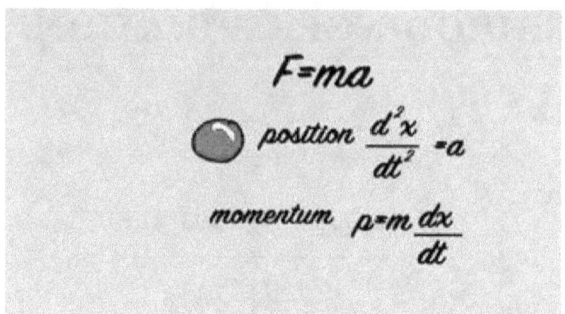

Once you have these two as shown in the figure above, you can derive basically everything else you could possibly know about it, velocity, energy etc. But when we get down to particle land newton's equations don't work anymore. If we put a particle in a box and we want to know where it is F=ma just doesn't cut it.

The Heisenberg uncertainty principle says that we can't know both the exact position and momentum of quantum objects, but, we can know other things, like the energy levels and the wave function, which we'll explore in this tutorial. But that information is all inside the schrodinger equation and with some hardcore mathematics we can tease it out. So before I get started i should say that this is the time independent version meaning that it doesn't involve time. And if at any point throughout this tutorial you get lost, don't worry, even schrodinger didn't know exactly what his equation meant.

So let's say we have a quantum system: an electron in a box. We want to know everything we can about this electron so we can make predictions, where it might be, what energy level it might be at. These answers are all buried within the Schrodinger equation.

So first let's start with this as shown in the figure below,

the pitchfork thing. This is the greek letter psi and it stands for what's called the wave function. It tells you where the electron is likely to be. But not where it will be. See quantum objects are sneaky in that you can't predict exactly where they'll be until you measure them. You can only predict where they'll probably be. Say there's this kid you know. You've grown up with them in your whole life. They're locked in their room with their homework, a play station, and a bed. If you had to guess where they were you'd say there's about an 80% probability they're on the playstation, 19% probability they're in bed and a 1% probability they're doing their homework. When you open the door you'll know for sure. But you were able to make these predictions because you know it. What if you had to guess where this electron is? You don't know this electron. Well that's what the wave function tells us. It gives us the probabilities of where it's likely to be. But a big difference is that while the wave function is only in one place at a time, the electron is in a superposition of all possible places at the same time.

You may have heard the famous thought experiment Schrodinger's cat who's in a superposition of being dead and alive until the box is opened and it's forced to choose a state. It's the same deal here. The act of not knowing where the electron is allows its probability distribution to be spread out over a large space kind of like a wave. Different kinds of waves can represent different probabilities of where it's likely to be.

WAVE FUNCTION.

It's a function that describes the wave shape of probability distribution of the electron. And when you open the door and measure where it is, this wave probability cloud function collapses and the electron becomes a particle again. No wonder Schrodinger was confused. So that's what the pitchfork means, the wave function tells us where our electron is likely to be.

Now let's take a look at this E as shown in the figure below.

$$E\Psi(x) = \frac{-\hbar^2}{2m}$$

ENERGIES THE ELECTRON
IS ALLOWED TO HAVE

It represents the energies the electron is allowed to have. Now before i go into what that means i just want to point out that the way this equation is arranged as show in the figure below these values as shown in the figure below,

$$E\Psi(x) = \frac{-\hbar^2}{2m}\frac{d^2\Psi(x)}{dx^2} + V\Psi(x)$$

are the ones we're trying to solve for, so it's telling us that if you do all this stuff as shown in the figure below,

$$E\Psi(x) = \frac{-\hbar^2}{2m}\frac{d^2\Psi(x)}{dx^2} + V\Psi(x)$$

you will find out the energy levels of the wave functions of the electron! And if you know these two things as shown in the figure below,

$$E\Psi(x) = \frac{-\hbar^2}{2m}\frac{d^2\Psi(x)}{dx^2} + V\Psi(x)$$

you can derive everything else, you can possibly know about the particle, just like the position and momentum of the ball. But let's backtrack a sec, what do i mean when i say energy levels the electron is allowed to have? Like it's a grown electron it can have whatever energy levels it wants right? Well, no. In the regular world we see around is energy can go up and down in a smooth continuous way, but this isn't the case in the quantum world and the reason comes from the wave like nature of the probability distributions. Because our particle is inside the box it has a zero probability of being found in or outside the wall, so this means the wave function always needs to be zero there, otherwise there's some probability that the electron could be outside the box which we know it's not.

That means that the electron can only have certain frequencies associated with it. This frequency is allowed as the wave function is zero at both edges, as shown in the figure below.

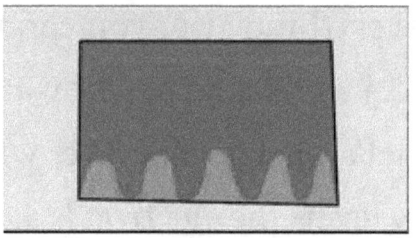

And this frequency is not as shown in the figure below,

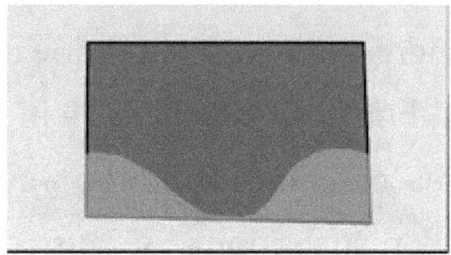

this frequency is allowed as shown in the figure below,

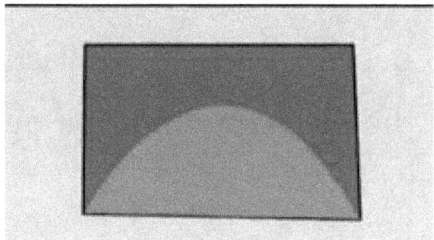

and this one is not as shown in the figure below,

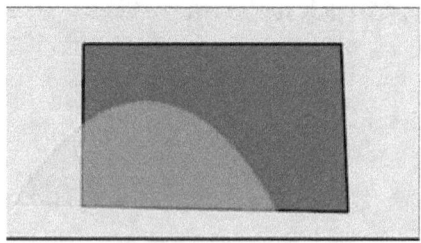

So Einstein discovered that energy is actually proportional to frequency by this relation E=hf where E is the energy, f is the frequency and this h here is planck's constant. As shown in the figure below,

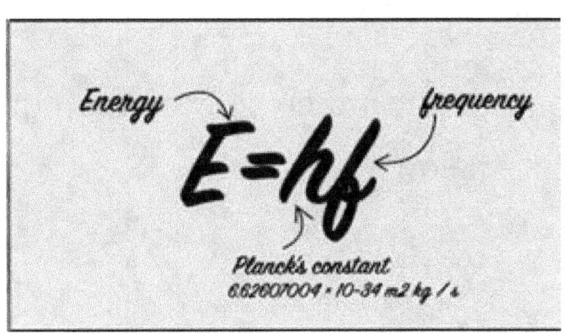

Don't worry too much about that h for now, all you need to know is that it's a constant which means its value doesn't change. So if only certain frequencies are allowed inside the box and this is a constant as shown in the figure below,

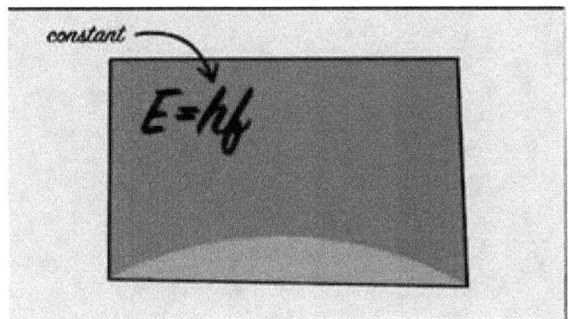

then it follows that only certain energy levels are allowed inside the box too.

This property of discrete or quantized values is where quantum mechanics gets its name. Things that can take on continuous values in the regular world like energy levels can only take on certain quantized values at the quantum scale.

Now let's look at the other side of the equation. As shown in figure below,

$$E\Psi(x) = \frac{-\hbar^2}{2m} \frac{d^2\Psi(x)}{dx^2} + V\Psi(x)$$

You know what you're solving for: the energy levels and the wave functions. But how is all of this going to help us get there? As shown in the figure below,

$$? \\ (x) = \frac{-\hbar^2}{2m} \frac{d^2\Psi(x)}{dx^2} + V\Psi(x)$$

Well overall energy is made up of kinetic energy and potential energy. As shown in the figure below,

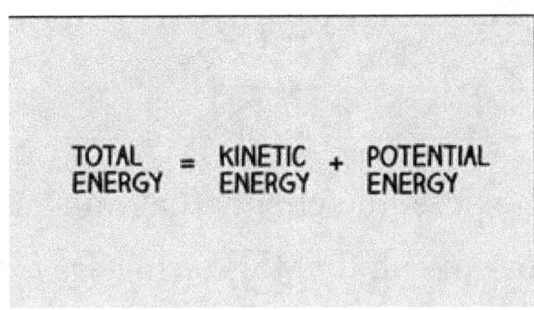

$$\text{TOTAL ENERGY} = \text{KINETIC ENERGY} + \text{POTENTIAL ENERGY}$$

If a skateboarder is on a skate ramp they'll be traveling at some speed and have some kinetic energy, but when they stop at the top they still have energy, it's just transformed into a different kind:

POTENTIAL ENERGY.

The entire energy of the system is just the kinetic energy plus the potential energy and while it can move back and forth between those two states its total value is always conserved. Sometimes potential energy is written as a V, so this term is the potential energy of the wave function as shown in the figure below.

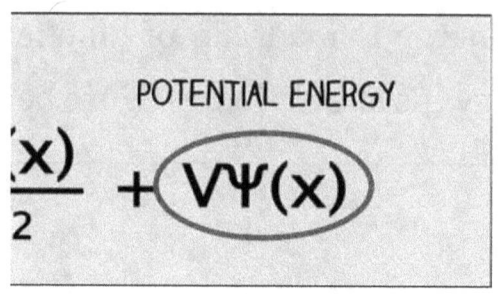

POTENTIAL ENERGY

$$\frac{x)}{2} + V\Psi(x)$$

So if this is the potential energy that must as shown in the figure above, mean that this term here is the kinetic energy as shown in the figure below.

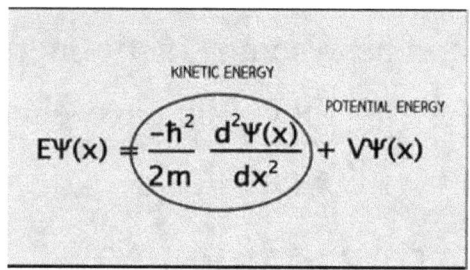

KINETIC ENERGY

POTENTIAL ENERGY

$$E\Psi(x) = \frac{-\hbar^2}{2m} \frac{d^2\Psi(x)}{dx^2} + V\Psi(x)$$

I know it doesn't really look like any kind of kinetic energy equation you've seen before right, so here's the derivation if you don't believe me. As shown in the figure below.

$$E = PE + KE$$

$$\therefore E = V + \frac{1}{2}mv^2 \boxed{\begin{array}{c} \text{momentum} \\ p = mv \end{array}}$$

$$\therefore E = V + \frac{p^2}{2m} \leftarrow \text{substitute}$$

$$\boxed{\psi = e^{i(kx-wt)}} \leftarrow \begin{array}{l} \text{general form of} \\ \text{wave equation} \end{array}$$

$$\frac{d\psi}{dx} = ike^{i(kx-wt)} = ik\psi$$

$$\frac{d^2\psi}{dx^2} = i^2 k^2 e^{i(kx-wt)} = -k^2\psi$$

$$\boxed{k = \frac{p}{\hbar}} \leftarrow \text{De Broglie Relation}$$

$$\therefore \frac{d^2\psi}{dx^2} = \frac{-p^2}{\hbar^2}\psi$$

$$-\hbar^2 \frac{d^2\psi}{dx^2} = p^2\psi$$

$$\text{and } E = \frac{p^2}{2m} + V$$

$$\therefore E\psi = \frac{p^2\psi}{2m} + V\psi$$

$$\therefore E\psi = \frac{-\hbar^2}{2m}\frac{d^2\psi}{dx^2} + V\psi \qquad \square$$

So if you can solve for the potential and kinetic energy of our quantum system, this will tell us the energy levels allowed, and that's everything there is to know about our little electron! So what would some typical Schrodinger solutions look like? Well to this particular problem all the solutions to the wave functions take these two forms as shown in the figure below.

SOLUTIONS

$$\psi(x) = \sqrt{\frac{2}{L}}\cos\left(\frac{\pi n x}{L}\right) \text{ when } n = 1, 3, 5\ldots \text{ (is odd)}$$

$$\psi(x) = \sqrt{\frac{2}{L}}\sin\left(\frac{\pi n x}{L}\right) \text{ when } n = 2, 4, 6\ldots \text{ (is even)}$$

And the energy equation that popped out was this as shown in the figure below.

$$E = \frac{\hbar^2 n^2 \pi^2}{2mL^2}$$

Well the first thing to note is that every term in this expression is either a constant or a whole number as shown in the figure above: h-bar is a constant, 2 is obviously a constant, m the mass of the electron is a constant, pi is a constant, and L the length of the box is a constant, and n stands for the different states of the electron and they're all whole numbers, 1, 2, 3, etc. So then the energy E can only have certain values as shown in the figure below.

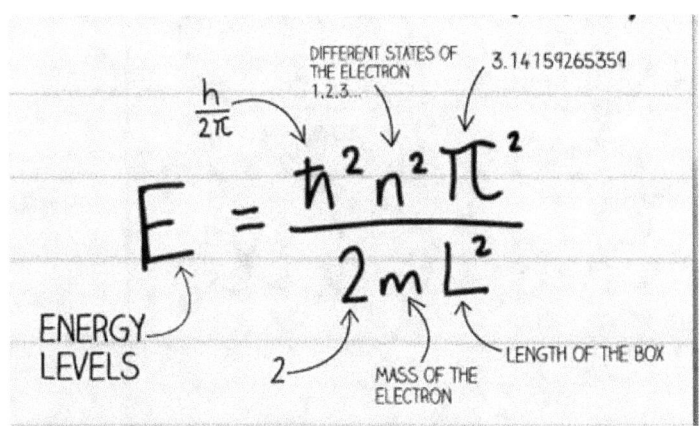

It's quantized. But what about the wave function?

Where is the electron? Well let's look at this as shown in the figure below.

$$\Psi(x) = \sqrt{\frac{2}{L}} \cos\left(\frac{\pi n x}{L}\right) \text{ when } n = 1, 3, 5\ldots \text{ (is odd)}$$

$$\Psi(x) = \sqrt{\frac{2}{L}} \sin\left(\frac{\pi n x}{L}\right) \text{ when } n = 2, 4, 6\ldots \text{ (is even)}$$

$$E = \frac{\hbar^2 n^2 \pi^2}{2m L^2}$$

when the electron is in its first energy state, when n is equal to 1. You get this as shown in the figure below.

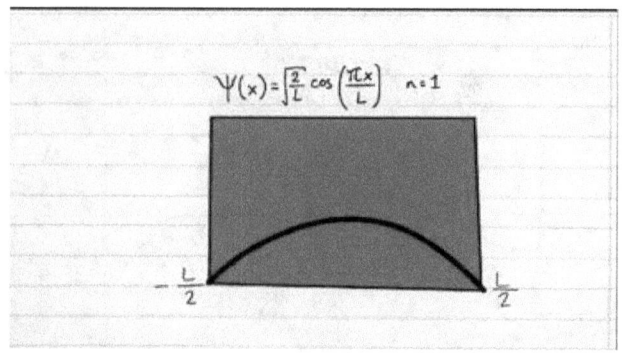

$$\Psi(x) = \sqrt{\tfrac{2}{L}} \cos\left(\tfrac{\pi x}{L}\right) \quad n=1$$

That's one of the wave functions of the electron and if you square it as shown in the figure below.

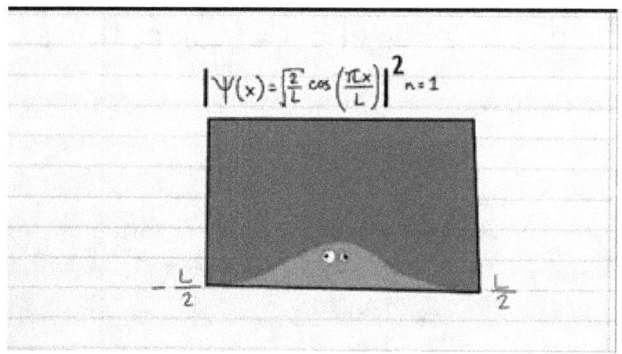

$$|\Psi(x)| = \sqrt{\tfrac{2}{L}} \cos\left(\tfrac{\pi x}{L}\right)|^2 \quad n=1$$

You get the probability distribution, a.k.a where the electron is hopeful to be. We can see that there's a high probability that it'll be found in the middle here as shown in the figure below, but a zero probability it'll be found right at the edges as shown in the figure below.

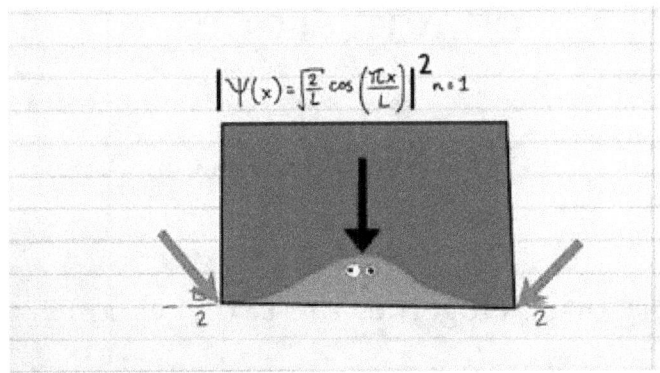

$$\left|\Psi(x) = \sqrt{\tfrac{2}{L}} \cos\left(\tfrac{\pi x}{L}\right)\right|^2_{n=1}$$

Here are some more wave functions and probability densities for other energy states. See how the wave function is always 0 right at the edges, as shown in the figure below.

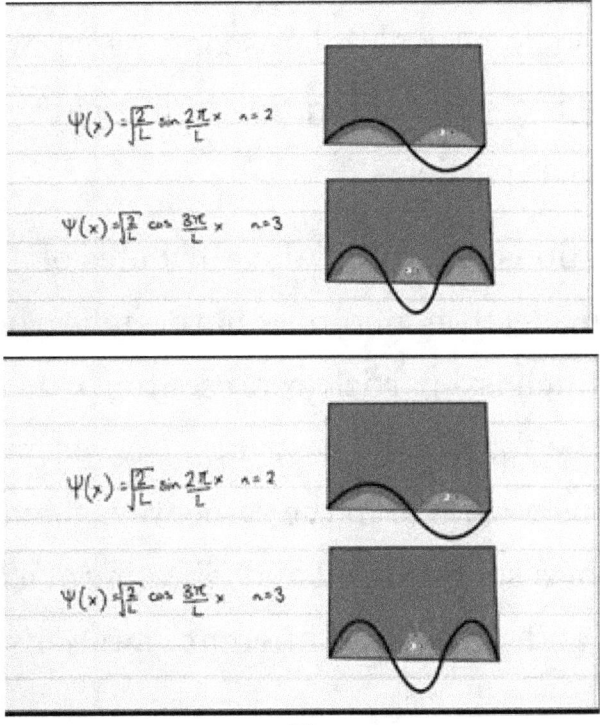

$$\Psi(x) = \sqrt{\tfrac{2}{L}} \sin\tfrac{2\pi}{L}x \quad n=2$$

$$\Psi(x) = \sqrt{\tfrac{2}{L}} \cos\tfrac{3\pi}{L}x \quad n=3$$

$$\Psi(x) = \sqrt{\tfrac{2}{L}} \sin\tfrac{2\pi}{L}x \quad n=2$$

$$\Psi(x) = \sqrt{\tfrac{2}{L}} \cos\tfrac{3\pi}{L}x \quad n=3$$

CHAPTER FOUR.

STRING THEORY.

This theory brings together the two backbones of 20th Century physics, quantum mechanics and albert einstein's theory of relativity, into a single framework that can describe all of the reality of physics. At its core, the theory is very simple. It suggests a particle isn't a point like object but an extended object, just like a string. It suggests that if we took any particle (electron, proton, quark, etc) and zoomed really close in on it with an extremely powerful microscope, what we would actually see is a string, vibrating in intricate ways that appear to us as particles.

These vibrations determine the attributes of a particle, be it proton, electron, photon, quark, etc. A photon, for example, which is a particle representing a quantum of light, may be formed by a string of a specific length striking a specific note. In the same way, a quark could be formed by a string folded and vibrating at a different frequency, and so on. The different vibrations are also what create the forces. What's even crazier is that while we can move in three different ways, up and down, left and right, and back and forth, a string can move in ten ways.

Strings are one-dimensional themselves, like a line, but they can move through multiple dimensions. And how they move through those dimensions is thought to be what gives things in the universe their attributes. It's kind like a guitar string. Strings on a guitar are one-dimensional, but they can vibrate through three dimensions, and how they vibrate determines the sound they make in the same way that in string theory a string vibration can produce a photon, a quark, and other particles.

➢ String Theory:

This was first introduced as a model of strong interactions, it was found that this theory could describe specific cases of particle scattering observed in the strong interactions between hadrons, the subatomic particles like the proton and neutron. Scientists, however, started to shift their focus from string theory after they realized that another theory called quantum chromo dynamics described the strong interaction much better.

This theory suggests subatomic particles called gluons are the ones that bind quarks together to make more familiar subatomic particles, such as protons and neutrons. However, the string theory is still strong and scientists use it to explain many things. The theory is one of the proposed methods for producing a theory of everything. Unlike the Standard Model of physics which can explain everything except gravity, string theory can explain everything, including gravity. The theory explains gravity the same way it explains other particles. A string can vibrate in a certain way that it can manifest itself as the hypothetical graviton, a quantum mechanical particle that carries the gravitational force. This theory is use by scientists to try to answer essential questions about the universe, such as what happens inside a black hole, how the big bang occurred, what is dark energy, the mysterious force that accelerates the expansion of the universe, etc. However, it's extremely hard to prove its predictions with the current technology, and many researchers have doubts whether it will ever work. But other researchers are confident that with the advancement of our technology, string theory will one day turn up results.

QUANTUM COMPUTING UNLEASHING THE POWER OF PARALLEL UNIVERSES.

➢ Quantum Computing:

This is an emerging field at the Forefront of technological advancement holds the potential to revolutionize the way we process information, by harnessing the principles of quantum mechanics, this cutting edge technology offers unparalleled computational power, far surpassing the capabilities of classical computers.

One of the Intriguing aspects of quantum computing lies in the concept of parallel universes, which allows for complex computations to be executed simultaneously. In this article we will explore the fundamental principles of quantum computing, and delve into the concept of parallel universes, uncovering how it contributes to the huge authority of quantum computing. Understanding quantum computing you to comprehend the concept of quantum computing, you need to grasp the basics of quantum mechanics, unlike classical bits in traditional computers quantum bits or qubits can exist in multiple states simultaneously, thanks to a phenomenon known as superposition.

➤ Superposition:

Quantum computers is allowed to process vast amounts of information in parallel by this property, leading to exponential computational speed up.

Furthermore, qubits can become entangled a phenomenon where the state of one qubit is inherently linked to the state of another. This entanglement enables quantum computers to perform complex computations by manipulating and measuring qubits collectively.

PARALLEL UNIVERSES IN QUANTUM COMPUTING.

Parallel universes also referred to as quantum parallelism is a concept that arises from the principles of superposition and entanglement in quantum mechanics. According to the many worlds interpretation proposed by physicist Hugh Everett, every time a quantum measurement occurs the universe splits into multiple branches with each branch corresponding to a different outcome of the measurement. This means that in the realm of quantum Computing, computations can occur simultaneously in multiple parallel universes it exploring different possibilities and outcomes.

The power of parallel universes, this lies in their ability to tackle complex computational problems more efficiently by simultaneously exploring various paths, quantum computers can potentially find solutions that would take classical computers an astronomical amount of time to compute. This parallelism opens up new avenues for solving optimization problems, simulating quantum systems and breaking encryption algorithms.

Real world applications, Industries such as pharmaceuticals, Finance, Logistics and cryptography stand to benefit significantly from the power of quantum Computing. For instance quantum computers could revolutionize drug discovery by simulating molecular interactions and accelerating the identification of potential drug candidates.

In the financial sector quantum algorithms can optimize investment portfolios and risk management strategies. It can also break traditional cryptographic systems, leading to the development of new encryption methods resistant to quantum attacks.

CHALLENGES.

- ➢ While quantum computing holds immense promise, several challenges need to be overcome before its wide spread adoption.
- ➢ Building stable and scalable quantum hardware is a significant hurdle as qubits are delicate and susceptible to de coherence.
- ➢ Additionally, developing error correction techniques and improving the reliability of quantum computations remain ongoing research areas. In the future advancements it may lead in various fields such as materials Science, artificial intelligence and climate modeling.

➢ The realization of fault-tolerant quantum computers could potentially revolutionize Industries and solve problems that are currently intractable.

CONCLUSION.

➢ Quantum computing represents a paradigm shift in computational power offering the ability to break complex problems exponentially faster than classical computers.

➢ Parallel universes stemming from the principles of superposition and entanglement enable it to explore multiple computational paths simultaneously. With the power to process vast amounts of information in parallel, quantum computing holds the key to unlocking ground, breaking advancements in various domains. As researchers and scientists continue to push the boundaries of this field we can anticipate a future where quantum Computing plays a pivotal role in transforming the way we solve complex problems and understand the universe

CHAPTER FIVE.

HEISENBERG UNCERTAINTY PRINCIPLE.

The heisenberg uncertainty principle is one of a handful of ideas from quantum physics to expand into general pop culture. This principle state that you can never contemporary know the right position and the right speed of an object and shows up as a metaphor in everything from literary criticism to sports commentary.

> ➢ Uncertainty:

This is often explained as a result of measurement that the act of measuring an object's position changes its speed or vice versa. More amazing and much deeper is the real origin, this principle occur because everything else in the universe behaves like both a particle and a wave at the same time.

In quantum mechanics the exact position and exact speed of an object have no meaning. To understand this you need to think about what it means to behave like a particle or a wave.

➤ Particles :

This exist in a single place at any instant in time. It be can represent by a graph showing the probability of finding the object at a particular place, which looks like a spike 100% at one specific position and zero everywhere else.

➤ Waves:

On the other hand our disturbances spread out in space like ripples covering the surface of a pond. you can clearly identify features of the wave pattern as a whole most importantly, its wavelength which is the distance between two neighboring peaks or two neighboring valleys, but you can't assign it a single position it has a good probability of being in lots of different places.

➤ Wavelength:

This is essential for quantum physics because an object's wavelength is related to its momentum mass times velocity, a fast moving object has lots of momentum which corresponds to a very short wavelength, a heavy object has lots of momentum even if it's not moving very fast which again means a very short wavelength, this is why you don't observe the wave nature of everyday objects. If you toss a baseball up in the air its wavelength is a billionth of a trillionth of a trillionth of a meter far too tiny to ever detect small things like atoms or electrons though can have wavelengths big enough to measure in physics experiments. so if you have a pure wave, you can measure its wavelength and thus its momentum but it has no position, you can know a particles position very well but it doesn't have a wavelength, so you don't know its momentum to get a particle with both position and momentum, you need to mix the two pictures to make a graph that has waves but only in a small area. How you can do this is by combining waves with different wavelengths which means giving our quantum object some possibility of having different momenta, when you add two waves, you find that there are places where the peaks line up making a bigger wave and other places where the peaks of one fill in the valleys of the other, the result has regions where you see that the waves are separated by regions of nothing at all, if you add a third wave in the regions where the waves cancel out get bigger a fourth hand they get bigger still with the wavy regions becoming narrower. if you keep adding waves, you can make a wave packet with a clearer wavelength in one small region that's a quantum object with both wave and particle nature, but to accomplish this, you had to lose certainty about both position and momentum, the position isn't restricted to a single point there's a good probability of finding it within some range of the center of the wave packet and you made the wave packet by adding lots of waves which means there's some probability of finding it with a momentum corresponding to any one of those, both position and momentum are now uncertain and the uncertainties are connected. if you want to reduce the position uncertainty by making a smaller wave packet, you need to add more waves which means a bigger momentum uncertainty, if you want to know the momentum better you need a bigger wave packet which means a bigger position uncertainty that's the heisenberg uncertainty principle first stated by german physicist werner heisenberg back in 1927.

This uncertainty isn't a matter of calculating well or badly but a fatal result of combining particle and wave nature, the uncertainty principle isn't just a practical limit on measurement it's a limit on what properties an object can have built into the fundamental structure of the universe itself.

QUANTUM GRAVITY.

➤ Quantum gravity:

This is a field of theoretical physics that seeks to define gravity based on the rule of quantum mechanics. This deals with situation in which neither gravitational nor quantum effects can be ignored such as in the vicinity of black holes or similar compact astrophysical objects such as neutron stars as well as in the early stages of the universe moments after the big bang.

Three of the four fundamental forces of nature are described within the framework of quantum mechanics and quantum field theory. They are,

> ➤ The electromagnetic interaction,
> ➤ The strong force and
> ➤ The weak force.

This leaves gravity as the only interaction that has not been fully accommodated. The current understanding of gravity is based on albert einstein's general theory of relativity which incorporates his theory of special relativity and deeply modifies the understanding of concepts like time and space.

Although general relativity is highly regarded for its elegance and accuracy, it has limitations.

The gravitational singularities inside of black holes, the ad hoc postulation of dark matter as well as dark energy and its relation to the cosmological constant are among the current unsolved mysteries regarding gravity. All of which signal the collapse of the general theory of relativity at different scales and highlight the need for a gravitational theory that goes into the quantum realm, at distances close to the plank length like those near the center of the black hole, quantum fluctuations of space, time are expected to play an important role.

The breakdown of general relativity at galactic and cosmological scales also points out the necessity for a more robust theory.

Finally, the discrepancies between the predicted value for the vacuum energy and The observed values which depending on the considerations can be of 60 or 120 orders of magnitude highlight the necessity for a quantum theory of gravity, the field of quantum gravity is actively developing and theorists are exploring a variety of approaches to the problem of quantum gravity, the most popular being m theory and loop quantum gravity. All of these upcoming goals is to describe the quantum action of the gravitational field which does not necessarily involve unifying all essential inter actions into a single mathematical framework.

However, many approaches to quantum gravity such as string theory try to develop a framework that describes all fundamental forces.

Such a theory is constantly assign as a theory of everything. Some of the approaches such as loop quantum gravity make no such attempt instead they make an effort to quantize the gravitational field while it is kept separate from the other forces.

Other lesser known but no less important theories include causal dynamical triangulation, non-commutative geometry and twister theory. One of the arduous of developing a quantum gravity theory is that direct observation of quantum gravitational effects is thought to only emerge at length scales near the plank scale around 10-35 m, a scale far smaller and hence only accessible with far higher energies than those currently available in high energy particle accelerators. Therefore, physicists lack experimental data which could differentiate between the disputing theories which have been considered. Thought experiment approaches have been suggested as a test testing tool for quantum gravity theories.

In the field of quantum gravity there are many open questions. Example it is not known how spin of elementary particles sources gravity and thought experiments could provide a pathway to explore possible resolutions to these questions even in the absence of lab experiments or physical observations. In the early 21st century, new experiment designs and technologies have arisen which suggest that indirect approaches to testing quantum gravity may be feasible over the next few decades.

This field of study is called phenomenological quantum gravity, the observation that all fundamental forces except gravity have one or more known messenger particles guide researchers to believe that at least one must exist for gravity. This hypothetical particle is known as the graviton. These particles act as a force particle similar to the photon of the electromagnetic interaction, under mild assumptions the structure of general relativity requires them to follow the quantum mechanical description of interacting theoretical spin two massless particles. Many of the accepted notions of a unified theory of physics since the 1970s assume and to some degree depend upon the existence of the graviton.

The Weinberg Whitten theorem places some constraints on theories in which the graviton is a composite particle. Gravitons are significance, theoretical step in a quantum mechanical description of gravity, they are generally believed to be undetectable because they interact too weakly. A conceptual difficulty in combining quantum mechanics with general relativity flow from the contrasting role of time within these two frameworks.

In quantum theories time acts as an independent background through which states evolve with the hamiltonian operator acting as the generator of infinitesimal translations of quantum States through time. In contrast general relativity treats time as a dynamical variable which relates directly with matter and moreover requires the hamiltonian constraint to vanish because this variability of time has been observed macroscopically, it removes any possibility of employing a fixed notion of time similar to the conception of time and in quantum theory, at the macroscopic level quantum gravitational effects are extremely weak and therefore difficult to test. For this purpose, the feasibility of experimentally testing quantum gravity had not received much attention prior to the late 1990s. However, in the past decade physicists have realized that evidence for quantum gravitational effects can guide the development of the theory. Since theoretical development has been slow the field of phenomenological quantum gravity which studies the possibility of experimental tests has obtained increased attention.

The most widely pursued potentiality for quantum gravity phenomenology involves gravitationally mediated entanglement, breach of Laurence invariance, imprints of quantum gravitational effects in the cosmic microwave background, in particular its polarization and de coherence induced by fluctuations in the space time foam. The latter scenario has been searched for in light from Gamay bursts and both astrophysical and atmospheric neutrinos placing limits on phenomenological quantum gravity parameters.

PHOTOELECTRIC EFFECT AND THE HISTORY OF EINSTEINS REVOLUTIONARY VIEW OF LIGHT.

The photoelectric effect is taught in nearly every physics class, but we often forget to tell why it's important. It's important, because einstein's photoelectric effect equation was based on a revolutionary new view of light. Now i use the word revolutionary, because einstein used that exact same term.

In april of 1901 when 22-year-old albert einstein read a paper by a conservative german physicist named max planck on how certain heated objects radiate light. Planck included this wild math trick, where he assumed that light was created in little bundles with an energy equal to a constant h, now called planck's constant times its frequency. Note, planck use the greek letter which looks like v for frequency. As shown in the figure below.

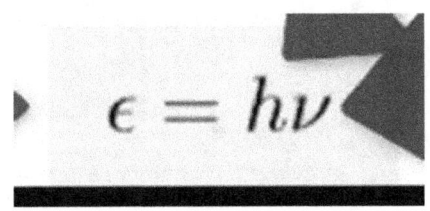

$$\epsilon = h\nu$$

Einstein was not impressed and wrote to his friend, about max planck studies on radiation, "misgivings of a fundamental nature have arisen in my mind, "so I'm reading his article with mixed feelings. A week later, einstein was still complaining that planck's idea that the thing that creates light are limited to how they can vibrate, an assumption i can't really warm to, "and because of planck, "my views on the nature of radiation have again sunk back into a sea of haziness. Einstein read a paper much more to his liking.

A former professor of mileva's, named phillip lenard, had written about an experiment that proved that shining ultraviolet light on a metal plate would cause electrons to fly off the plate. This was only a few years after the discovery that something called cathode rays in vacuum tubes were actually beams of electrons.

Photoelectric Effect

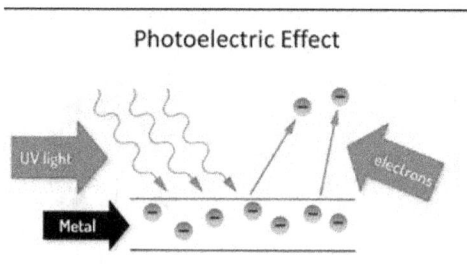

This phenomenon quickly became known as the photoelectric effect, as shown in the figure above, photo for light, and electric for electric.

Einstein wrote mileva, " kitten, just look at a wonderful paper by Lenard "on the generation of cathode rays by ultraviolet light. Under the control of this beautiful piece and i'm filled with such happiness and joy that i absolutely must share it with you." There's no hint in this letter that einstein saw any combination among the photoelectric effect and planck's energy bundles.

Einstein read that phillip lenard had conducted a new experiment on how ultraviolet light creates electrons and the photoelectric effect that einstein called pioneering. Lenard had the very clever idea of adding an extra voltage the wrong way, which he called a counter voltage. So that the electrons freed by the ultraviolet light would be pushed back towards the irradiated plate. As lenard said, in the same way as a stone thrown upward falls back to the ground. Lenard knew that the voltage to stop the electrons from reaching the other plate could tell him about the top energy of the freed electrons.

Then lenard found something truly startling. The top speed of the freed electrons was unaffected by the intensity of the light. Lenard determine that the energy of the electrons did not come from the light, alternatively, the light behaved like a trigger to release the inherent energy of the atoms like the fuse in decharging a loaded gun. This triggering theory seemed like a good remedy. And even as late as it was 1909, it was written as one of the generally accepted truths of physics.

However in 1904 or so, einstein had a quirky thought. What if the energy of the electrons did come from the light, but the light didn't act like a typical wave.

For a typical wave, like a sound wave or a water wave, the higher the intensity, the greater the energy. So more intensity ultraviolet light should lead to higher energy freed electrons, but that's not what lenard had observed. However, einstein remembered planck's paper were planck assumed that light was created in litte energy packets with an energy that depended on the frequency of light. Now, planck just did this as a math trick, but einstein started thinking, what if it was literally true? What if light was composed of tiny little particles that could only be created, moved or absorbed as a whole.

These particles is called quanta by einstein, which is a word for units that are indivisible, which is why this branch of physics is called quantum mechanics, quantum being the singular of quanta.

In 1905, einstein craft, the assumption to be contemplated here is that when a light ray is spreading from a point, the energy is not distributed continuously over ever-increasing spaces, but consists of finite number of energy quanta that are localized in points in space, move without dividing, and can be absorbed or generated only as a whole.

This has been called, the most revolutionary sentence written by a physicist of the 20th century. Before this paper, items were either waves or particles. This was the first proposal that they could be both. Light has frequency and creates interference patterns like a wave, but is composed of little energy packets that cannot be split in two just like you can't split an electron in two like a particle. Later, einstein described it, as a kind of fusion of the wave and particle theories of light. And in 1951, he wrote his friend Besso, all these 50 years of pondering have not brought me any closer to answering the question,

WHAT ARE LIGHT QUANTA.

What are light quanta?" Despite not really getting what light quanta are, einstein found that this counterintuitive idea of treating waves like a bunch of particles could explain all sorts of phenomenon, including the photoelectric effect.

Einstein decided in the photoelectric effect, electrons on the plate are stuck to the metal by electric forces. And it can only be removed by giving them a certain amount of energy, which they get from absorbing the energy of the incoming light. Einstein then postulated that the light came in a little energy packets, where the brighter the light, the more packets, that the individual light packets, now called photons, all of them have equivalent energy. If the frequency is too low, the photons of light have less energy than the binding energy of the surface electrons to the metal, and no electrons will be freed from the plate no matter how many packets you throw at the plate. This is why you need relatively high frequency light to create the photoelectric effect, also why UV light creates sunburn and visible light does not. Above this threshold, the more intense the light, the more electrons are freed, but all the electrons have the same maximum energy, the energy of the photon of light minus the minimum energy without charge of the electron.

According to planck's equation, the energy of light is planck's constant h times the frequency as shown in the figure below.

$$KE_{max} = \text{Energy of light} - \text{Binding Energy}$$

$$hf$$

And according to einstein, the minimum energy to free the metal is a constant of the metal as shown in the figure below.

$$KE_{max} = \text{Energy of light} - \text{Binding Energy}$$

$$hf \quad - W$$

Einstein also knew that from energy conservation, that the voltage to stop all the electrons from moving to the other plate times the charge on the electron equals to the maximum kinetic energy of the electron.as shown in the figure below.

$$KE_{max} = \text{Energy of light} - \text{Binding Energy}$$

$$qV = hf \quad - W$$

Thus, einstein created the equation for the photoelectric effect as shown in the figure above. In this equation, for any metal, there are only two variables in this equation: the stopping voltage and the frequency. Therefore, einstein theorized that if you graph the stopping voltage versus the frequency of light, you should get a straight line no matter what material you use for the plate. Even more impressively, the slope of that line can give you planck's constant h. as shown in the figure below.

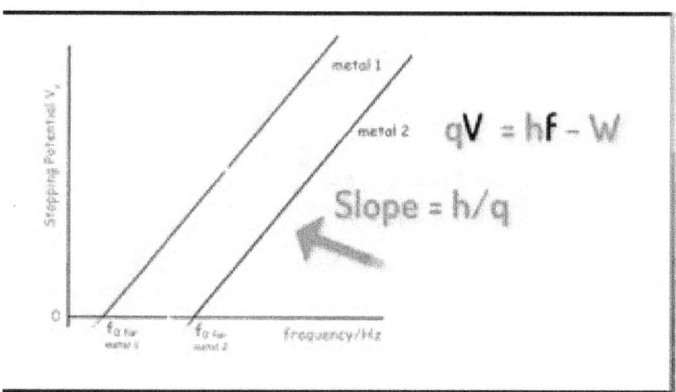

Einstein found an experimental way to prove that light that comes in little energy packets with an equation equal to a constant times the frequency of light, and even found a new way to find a universal constant of nature.